ASTRONAUT COLORING BOOK

BY KID KONGO

ISBN: 10: 1532755856
ISBN-13: 978-1532755859

I0482146

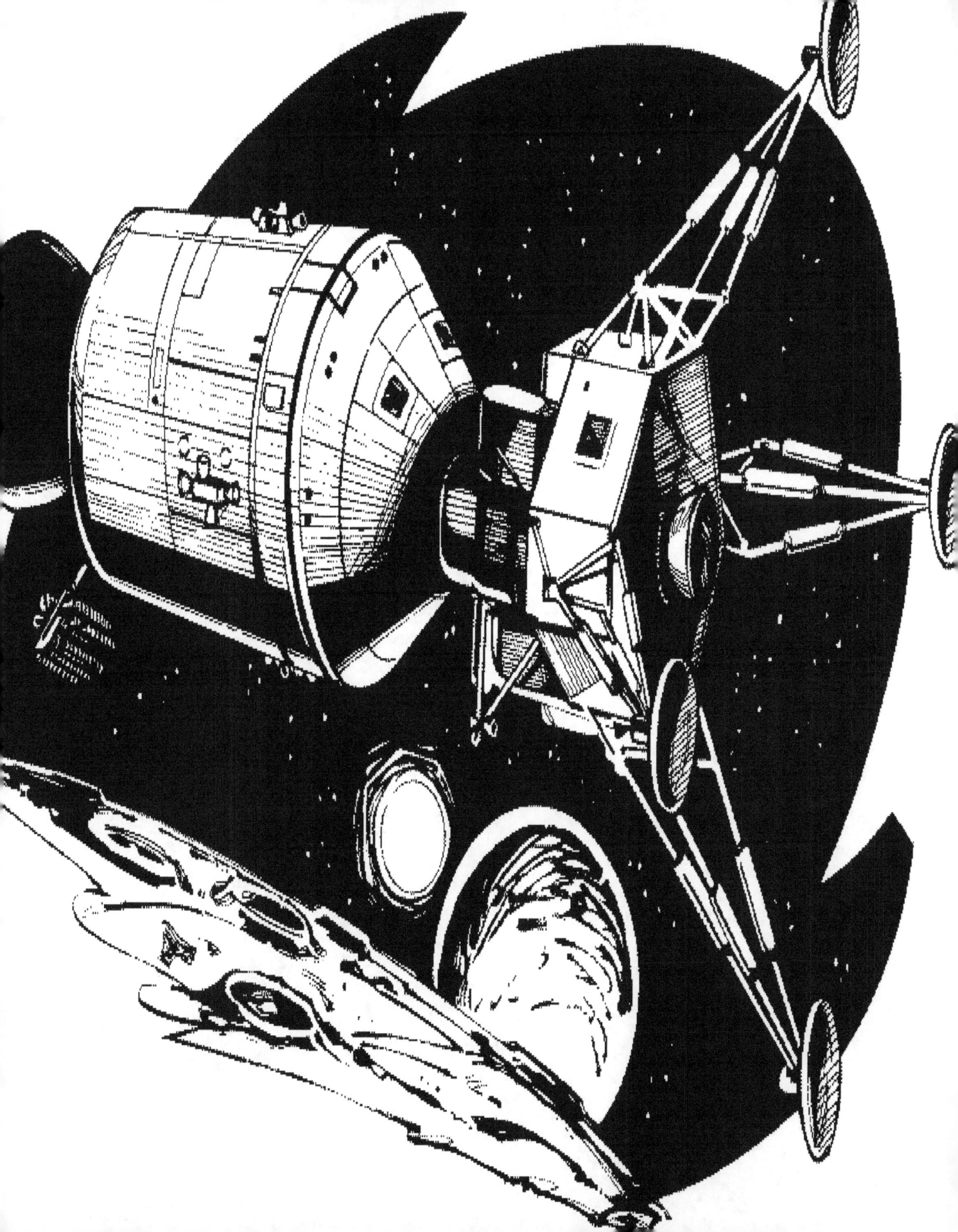

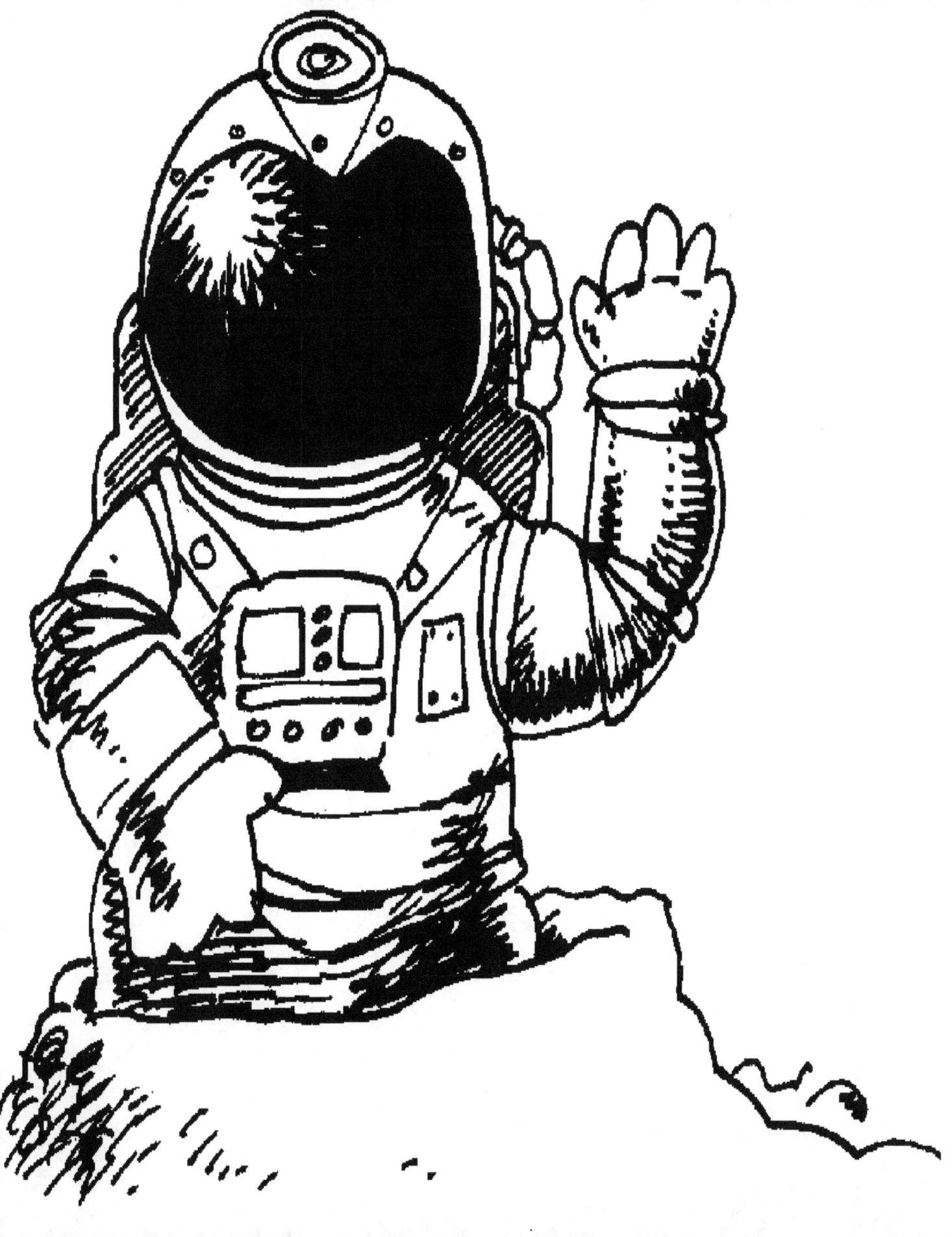

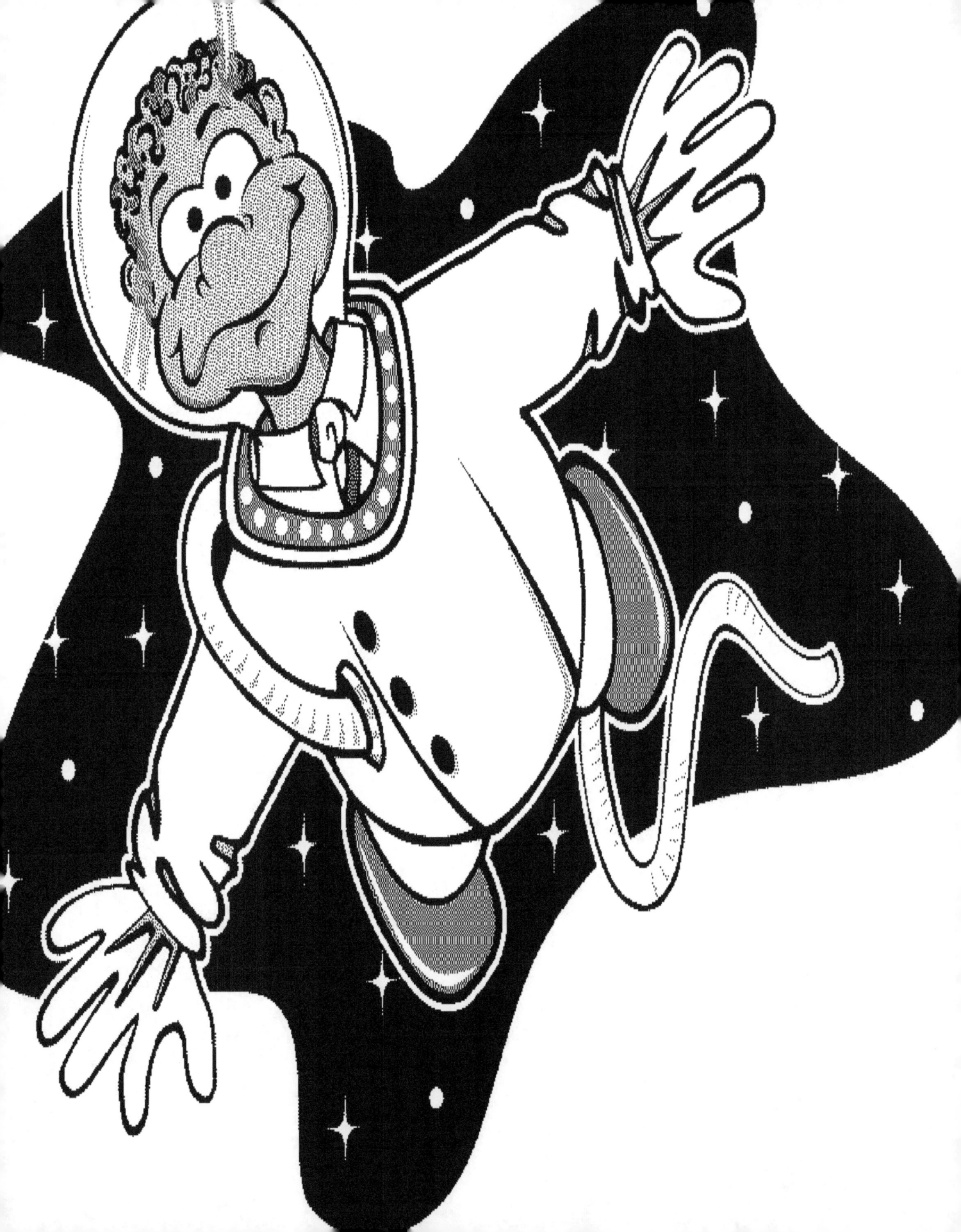

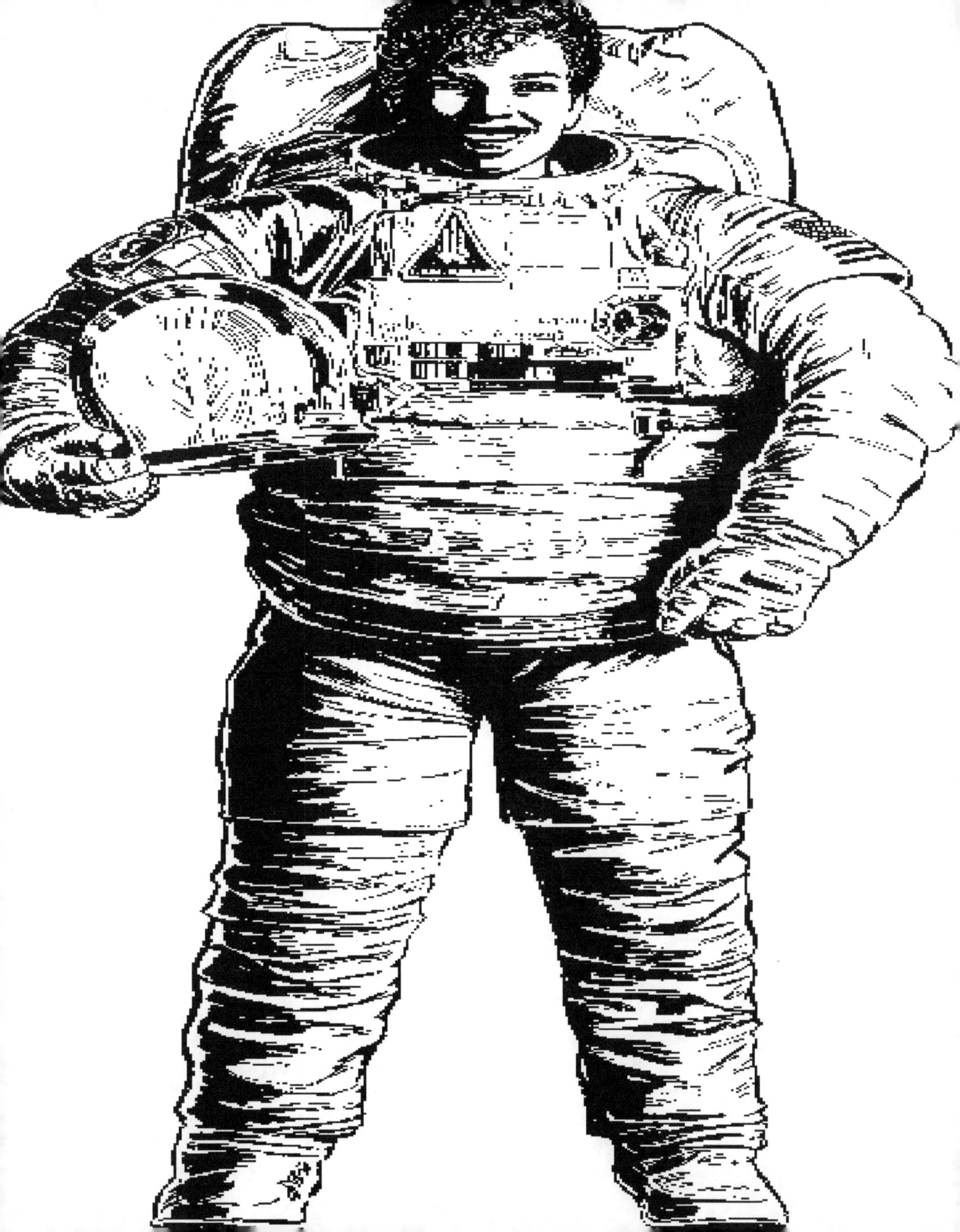

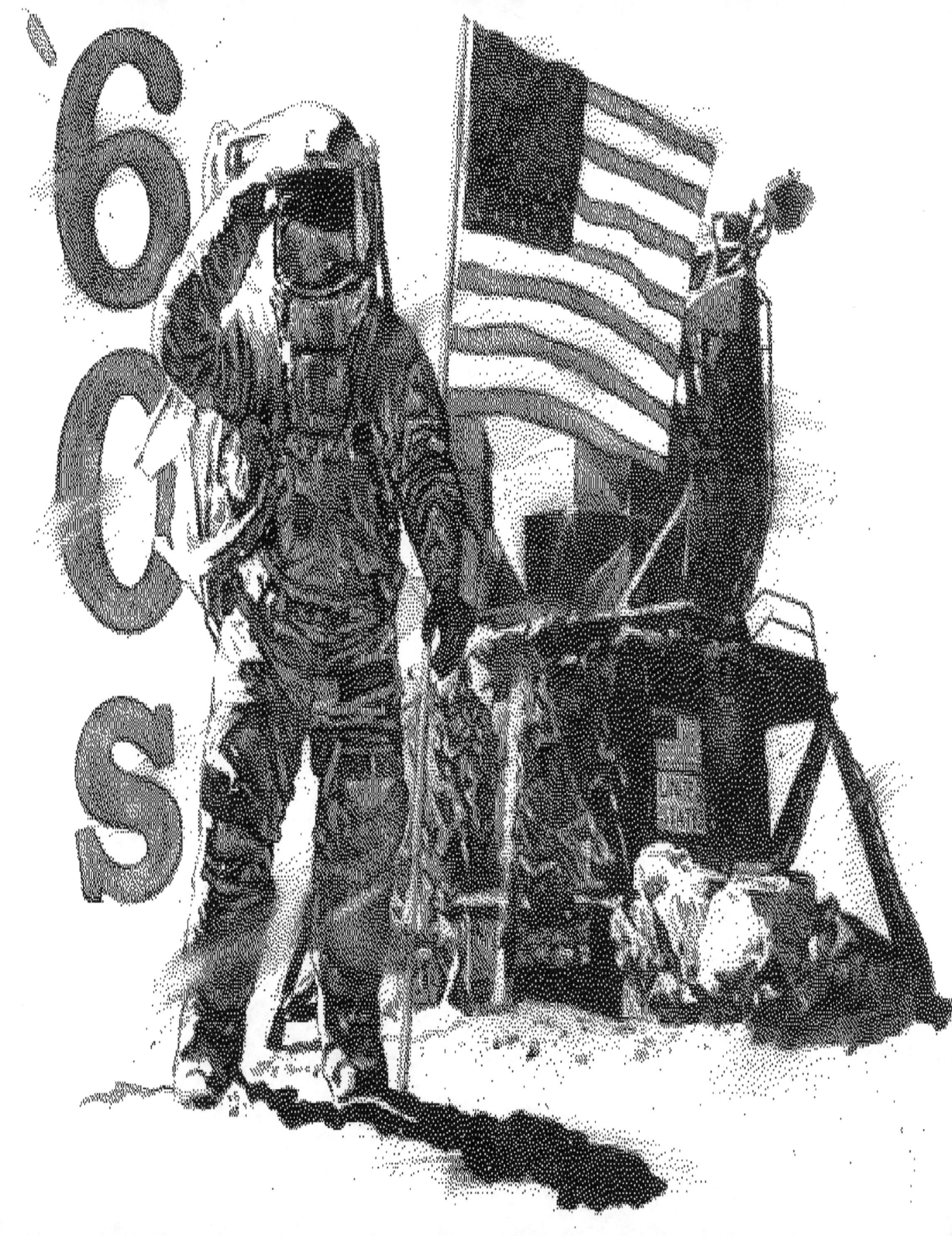

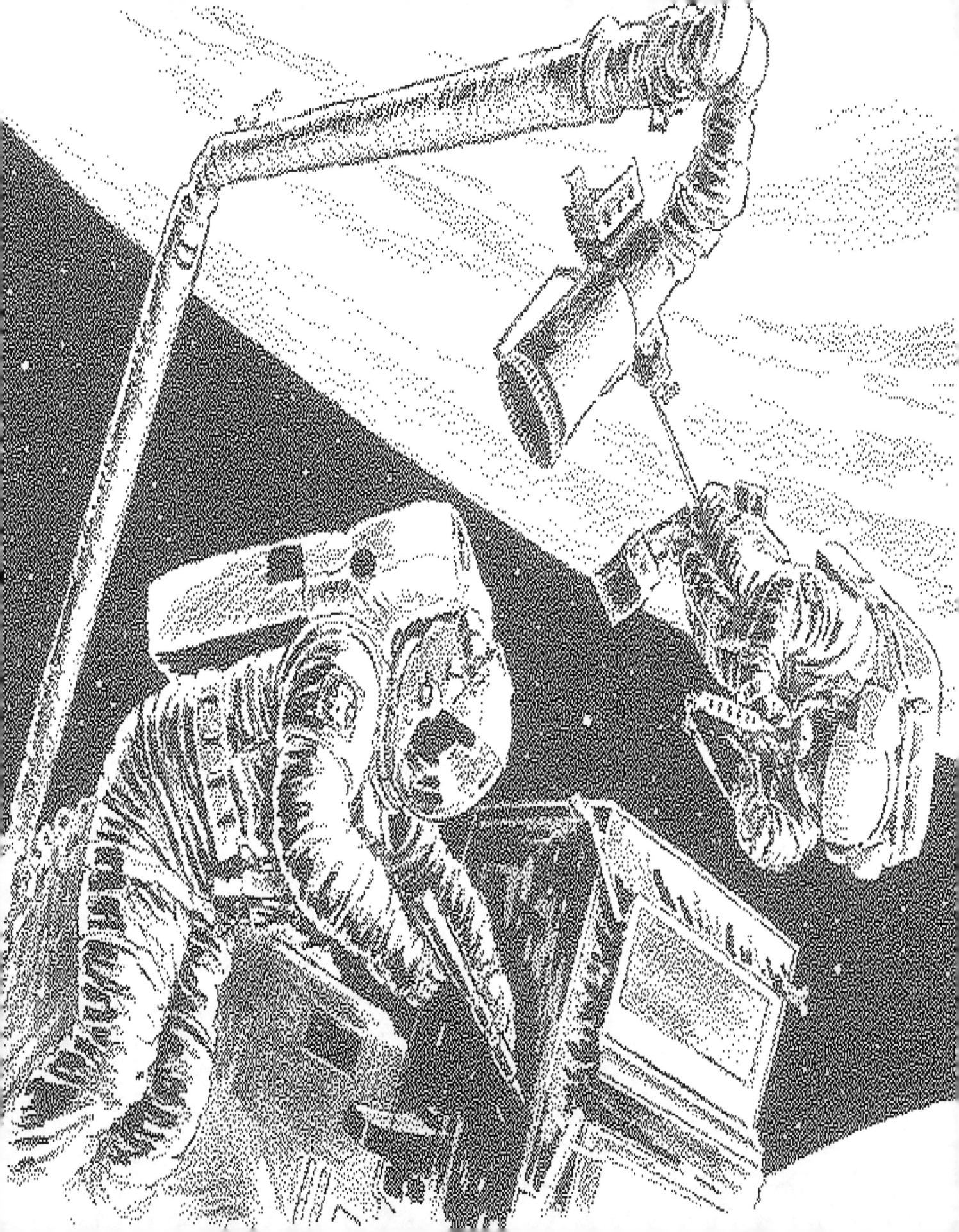

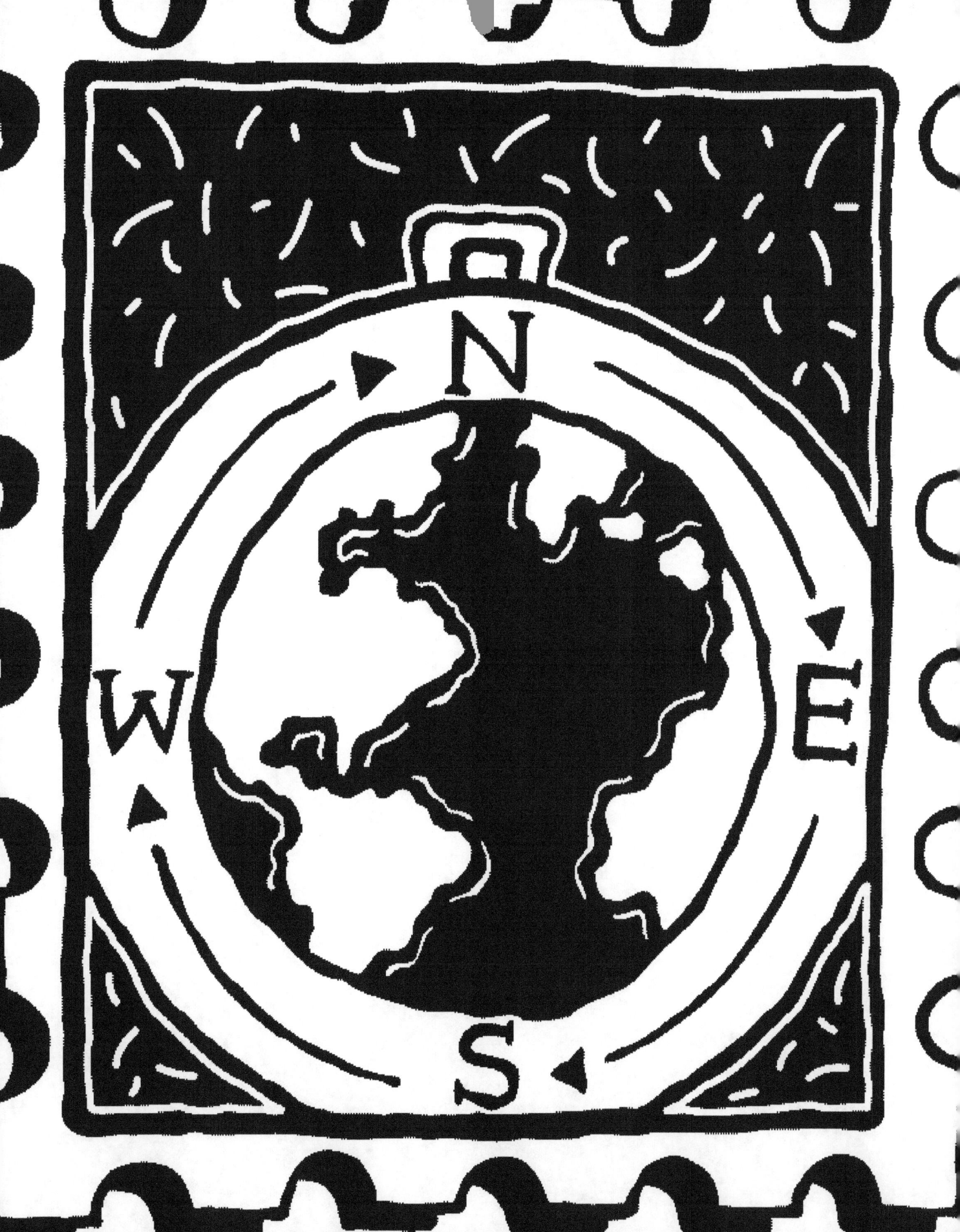

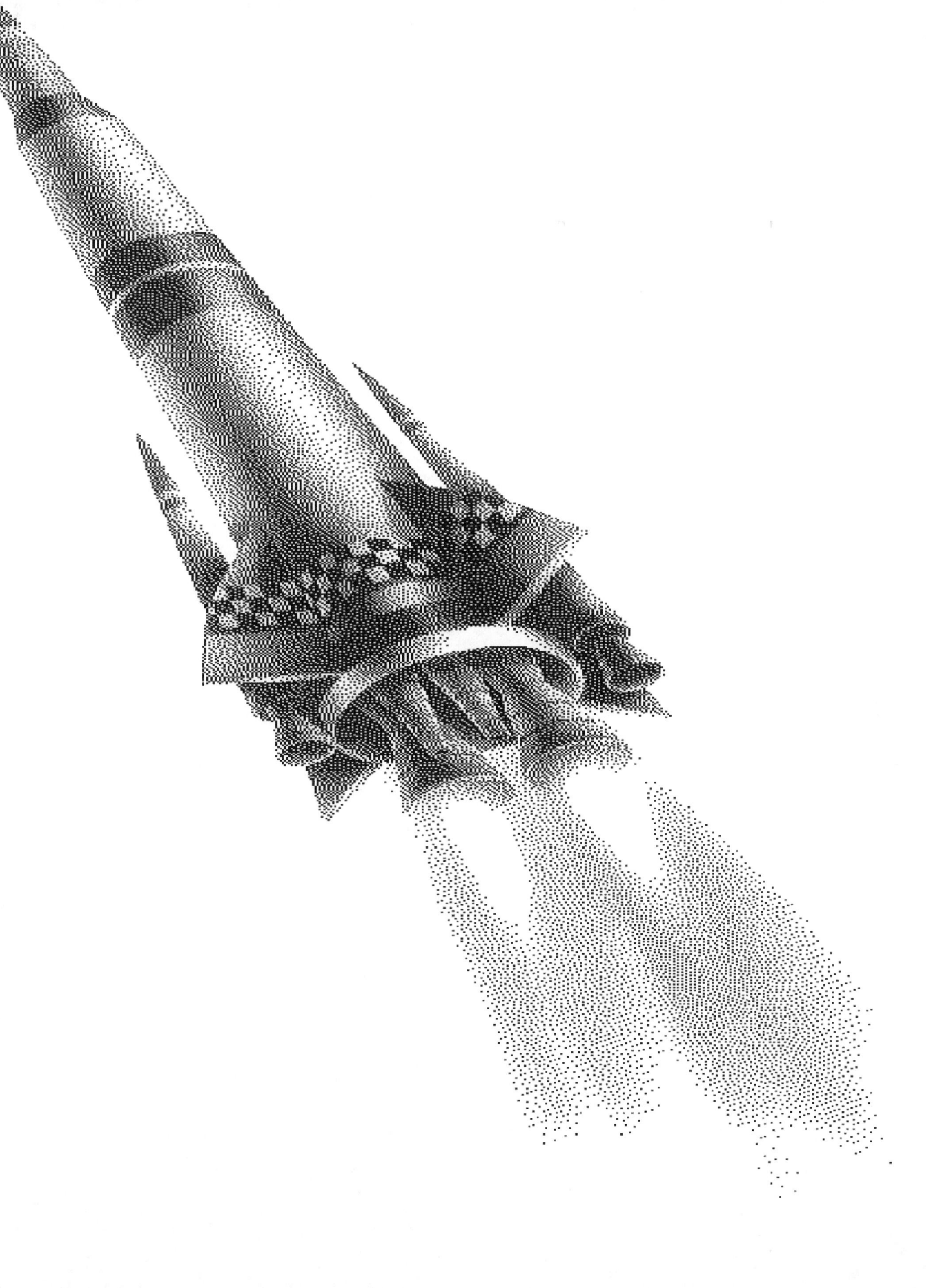

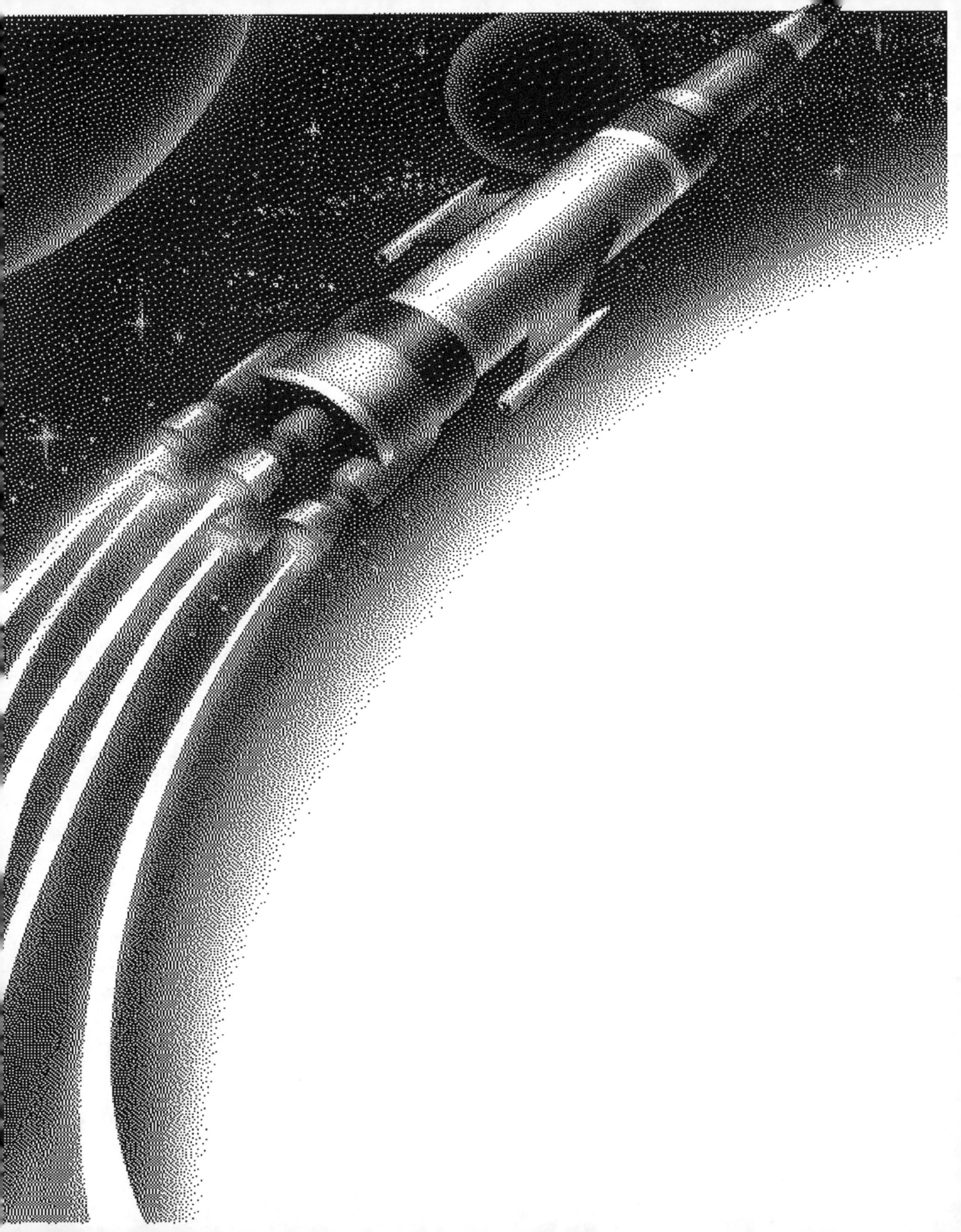

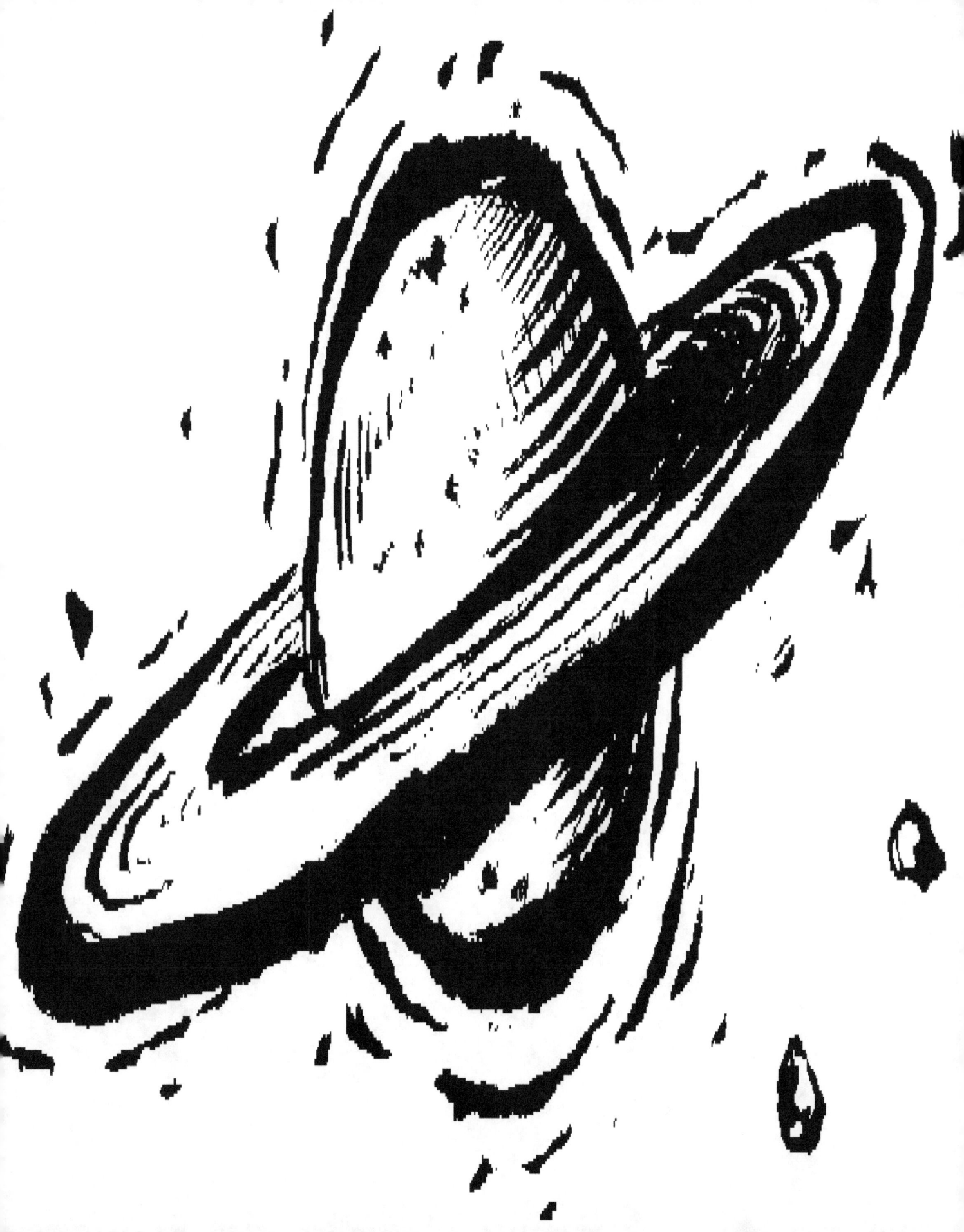

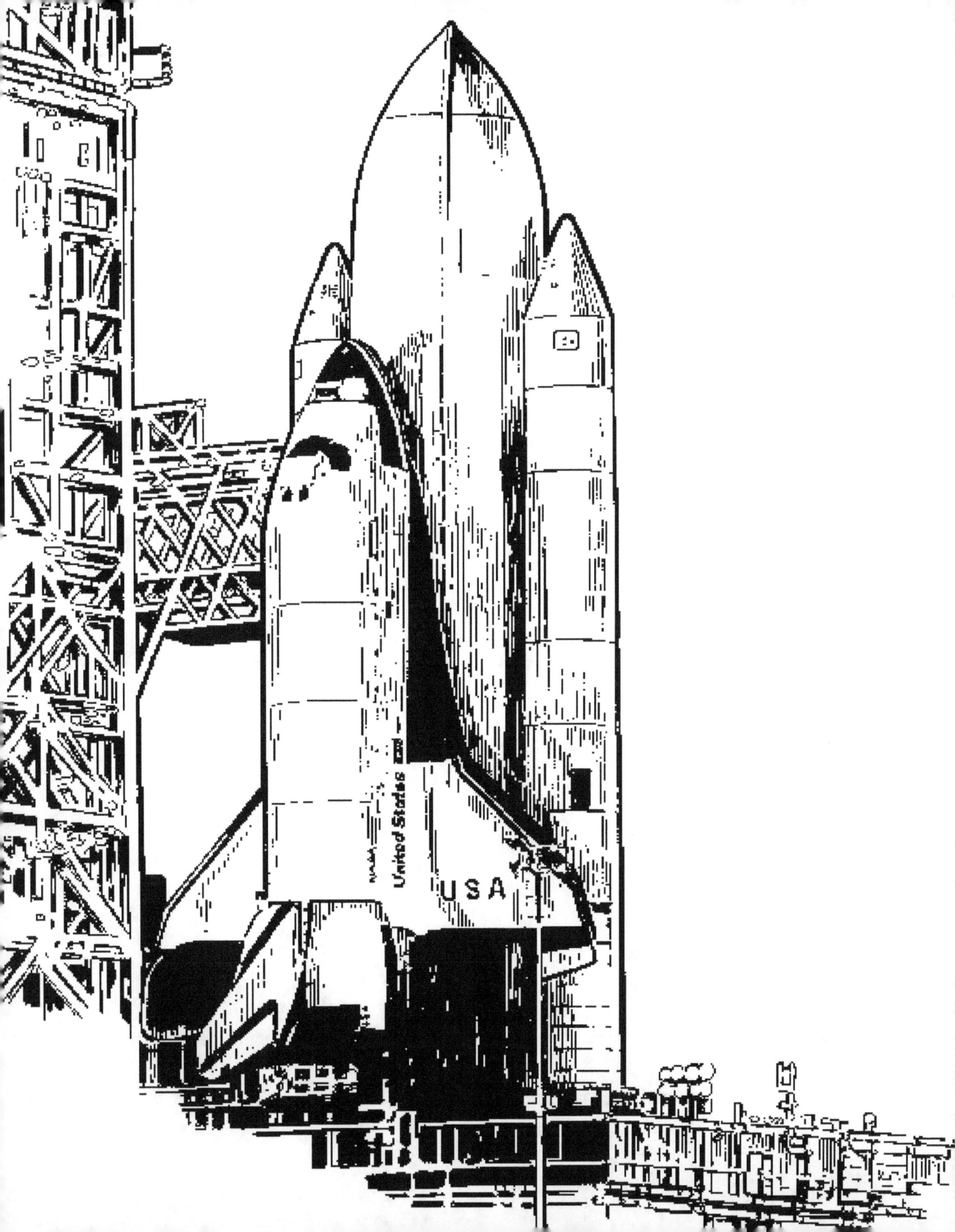

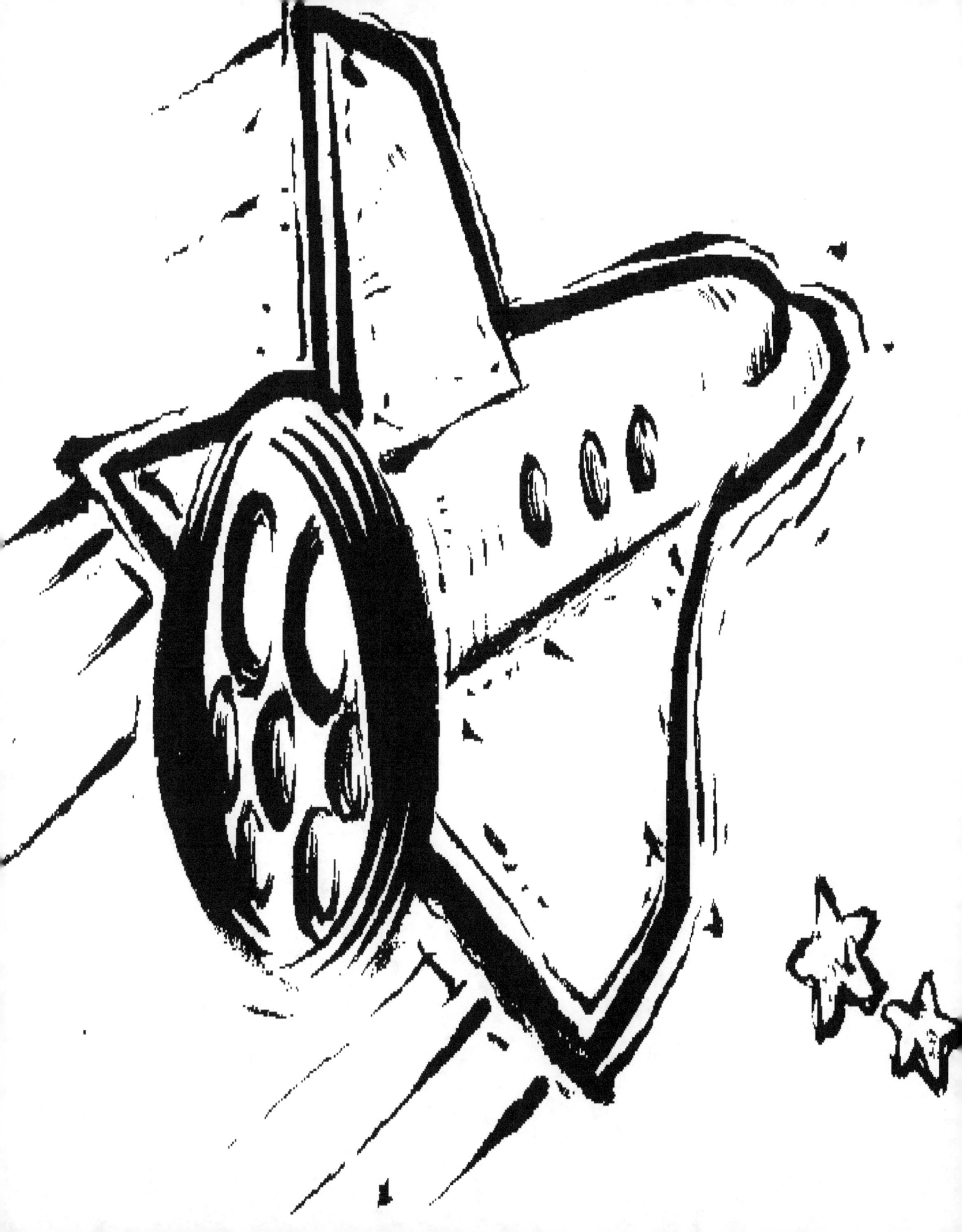

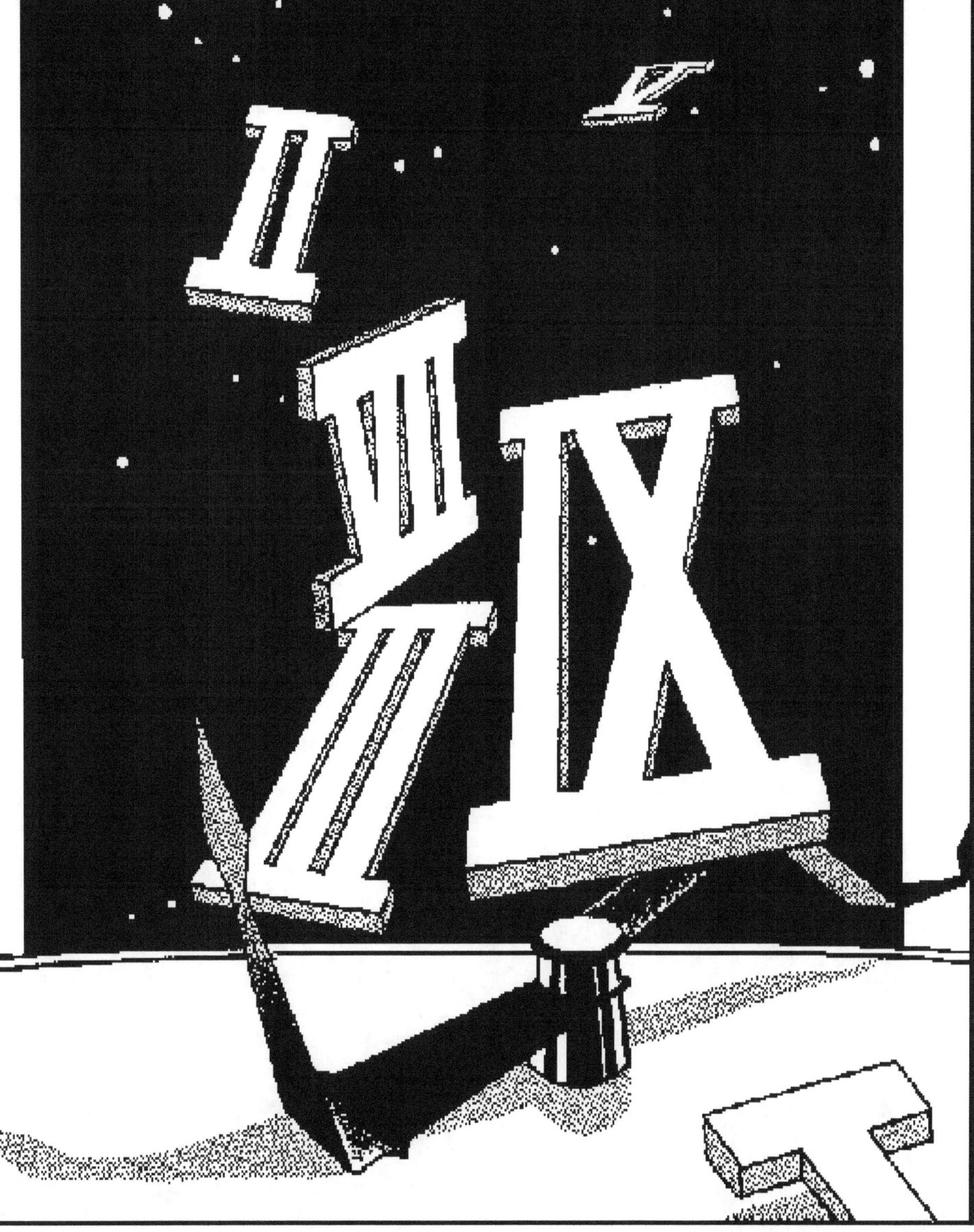

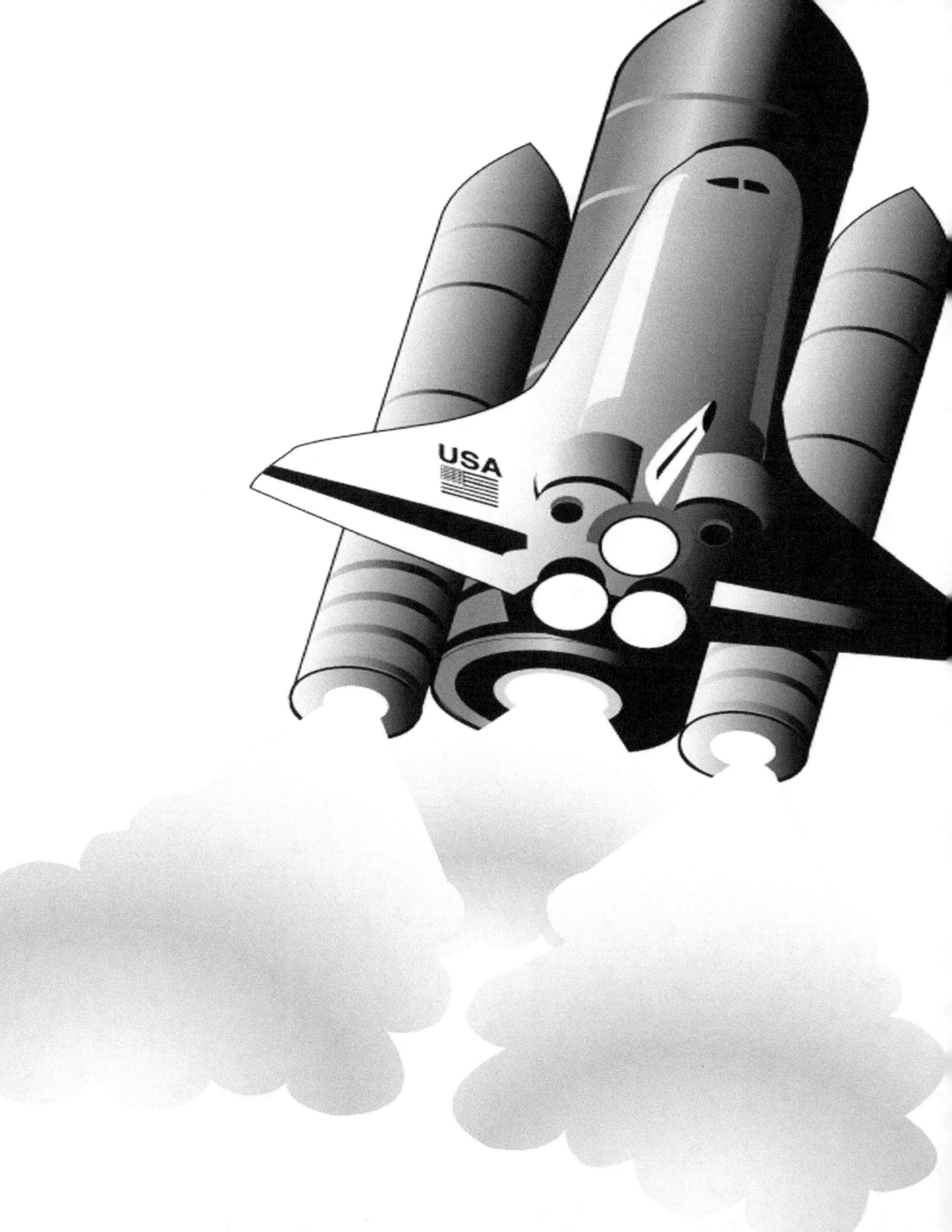

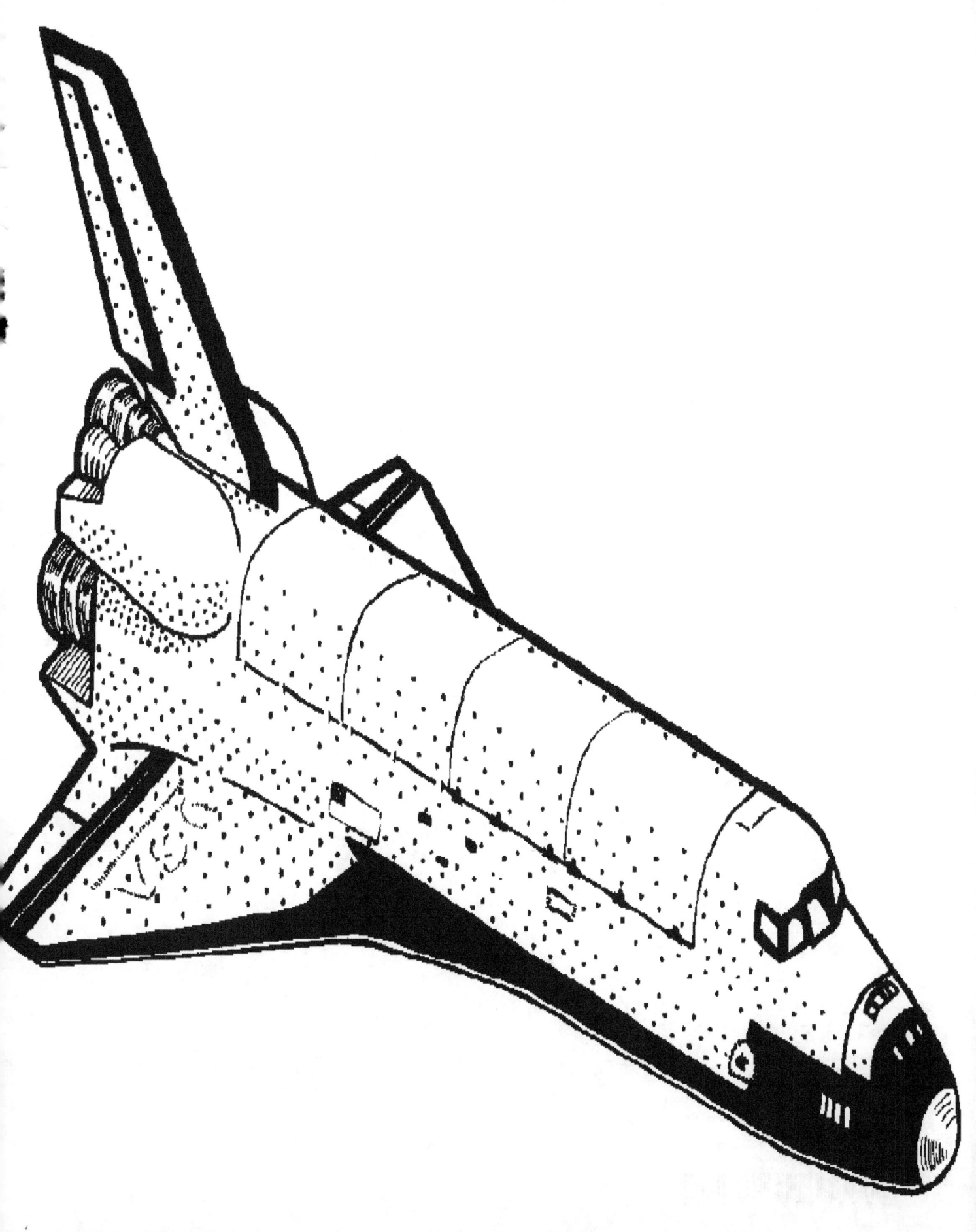

www.ingramcontent.com/pod-product-compliance
Lightning Source LLC
Chambersburg PA
CBHW080528190526
45169CB00008B/3099